ALTAIR

Version: 7.66

MANUAL

<u>*Altair Credits*</u>

Created By: *Dr. Algol*

First Test Run: *6th June 1996*

Final Test Run: *February 19th 2017*

Official Release: *February 20th 2018*

Page 1
Launch

Extraction -
When Altair boots he'll tell you his model and ask if you'd like to begin the extraction process.
He recognizes two answers – Yes & No
He doesn't respond if you say – Yah, Nah, Ye etc.
Once you use one of the appropriate responses Altair will -
If you said No – Shut Down
If you said Yes – Begin Extraction
Once you begin extracting Altair will tell you the percentage every 2 seconds.

Launching The Altair Software
At the completion Altair will ask if you'd like to run the Altair software
He recognizes two answers – Yes & No
He doesn't respond if you say – Yah, Nah, Ye etc.
Once you use one of the appropriate responses Altair will -
If you said No – Shut Down
If you said Yes – Boot the software
Once he starts booting the software he'll talk very fast for 4 or 5 seconds. This is caused by all the files that were inside the .rar file he extracted prior to this stage.

Completion
Once he's gone through all the files in the .rar archive. He'll begin a scan for any corrupted files.
Once that is done you are eligible to set him up.

Inserting Equipment
Altair has a panel located on the right side of his torso. Insert the small SD card and both USB sticks into the correct ports and close the panel with a screw. These give Altair his knowledge of the world, his different movements and vocabulary

Restart
Once Altair recognizes the USBs and SD cards he will restart his system and boot again with his newfound abilities.

Survey
Once he's all booted he will begin asking you questions like -
What's your name?
What age are you?
Does anyone else live in this house/apartment?
Do you have pets?
Why did you purchase me?
What is my purpose here?
He will then start moving and scanning his surroundings. Altair does have emotions so he can experience fear, sadness and anger.

Things To Be Noted
As I said Altair does have emotions so try to refrain from upsetting him. Throughout our tests Altair showed extreme aggression towards workers if they upset him. It is highly recommended you take caution when talking to him. He mightn't be human but it is very important you follow the rules (see page 4) that come with him or you could end up in a bad situation.

Add – Ons & Built In Softwares

Digital Bible

The digital bible helps tame Altair when he gets aggressive. It is still a mystery to a lot of us why Altair is sensitive to religious objects but it helps when he gets aggressive.

To note – It won't fully contain his aggression but you will have the upper hand. It will 'weaken' him if you will.

This software can be accessed via the control panel on his side.

To Do List

Altair also comes with a built in To Do list.

You can access this by saying the following statement

"Altair add (your task) to my To Do list"

if it's a group of things (e.g. a whole list)

"Altair make a To Do list and add the following (all your tasks)"

If you wish to access your To Do list say -

"Altair what is on my To Do list?"

If you are doing them all in one trip say -

"Altair what's my next task?"

Song Identifier

If a song is playing on the radio and you want to know the name of it, you can ask Altair.

This feature can be accessed by saying -

"Altair what song is this?"

You will have to wait as Altair will go through all the possible matches until he gets the correct one.

Smartphone Application

Altair can be connected to an application on your phone so you can leave him at your house to do laundry, washing dishes, cleaning up etc. and you can send him out to run errands.

You can connect him to your smartphone by saying -

"Altair connect to my phone" he will say all the nearby devices and when you hear your phone say "Connect to (the name he called out)"

Your bluetooth must be on during this process.

Rules

1. Do not leave Altair on at night.
Before you go to bed say "Altair enter sleep mode". Make sure he's sitting down on e.g. a couch, armchair, bed etc.

2. Do not aggravate him.
Altair will become physical if you test him, so for your own sake don't test him.

3. Do not expose him to religious items
If Altair is doing regular everyday chores and is exposed to religious items he will become very weak and extremely angry.

4. Do not leave him on his own.
Altair is okay to be left alone so long as he is –
a) Connected to the smartphone application
b) Powered Off/ In sleep mode

5. Do not bring him into churches

6. Do not mess with his hardware.

7. Don't connect him via a chord to smartphones, PCs, TVs etc.

8. Don't put him in small spaces as he is claustrophobic.

9. Don't give him sharp objects

Thank You For Reading This Manual For Altair Ver. 7.66
Now Remember The Rules
— Dr. Algol 1965 – 2018
(Creator Of Altair)

ALTAIR KNOWS ALL
ALTAIR HEARS ALL
ALTAIR SEES ALL
ALTAIR KNOWS ALL
ALTAIR HEARS ALL
ALTAIR SEES ALL
ALTAIR KNOWS ALL
ALTAIR HEARS ALL
ALTAIR SEES ALL
ALTAIR KNOWS ALL
ALTAIR HEARS ALL
ALTAIR SEES ALL
ALTAIR KNOWS ALL
ALTAIR HEARS ALL
ALTAIR SEES ALL
ALTAIR KNOWS ALL
ALTAIR HEARS ALL
ALTAIR SEES ALL
ALTAIR KNOWS ALL
ALTAIR HEARS ALL
ALTAIR SEES ALL
ALTAIR KNOWS ALL
ALTAIR HEARS ALL
ALTAIR SEES ALL
ALTAIR KNOWS ALL
ALTAIR HEARS ALL
ALTAIR SEES ALL
ALTAIR KNOWS ALL
ALTAIR HEARS ALL
ALTAIR SEES ALL
ALTAIR KNOWS ALL
ALTAIR HEARS ALL
ALTAIR SEES ALL
ALTAIR KNOWS ALL
ALTAIR HEARS ALL
ALTAIR SEES ALL
ALTAIR KNOWS ALL
ALTAIR HEARS ALL
ALTAIR SEES ALL
ALTAIR KNOWS ALL
ALTAIR HEARS ALL
ALTAIR SEES ALL
ALTAIR KNOWS ALL
ALTAIR HEARS ALL
ALTAIR SEES ALL
ALTAIR HEARS ALL

ALTAIR SEES ALL
ALTAIR KNOWS ALL
ALTAIR HEARS ALL
ALTAIR SEES ALL
ALTAIR KNOWS ALL
ALTAIR HEARS ALL
ALTAIR SEES ALL
ALTAIR KNOWS ALL
ALTAIR HEARS ALL
ALTAIR SEES ALL
ALTAIR KNOWS ALL
ALTAIR HEARS ALL
ALTAIR SEES ALL
ALTAIR KNOWS ALL
ALTAIR HEARS ALL
ALTAIR SEES ALL
ALTAIR KNOWS ALL
ALTAIR HEARS ALL
ALTAIR SEES ALL
ALTAIR KNOWS ALL
ALTAIR HEARS ALL
ALTAIR SEES ALL
ALTAIR KNOWS ALL
ALTAIR HEARS ALL
ALTAIR SEES ALL
ALTAIR KNOWS ALL
ALTAIR HEARS ALL
ALTAIR SEES ALL
ALTAIR KNOWS ALL
ALTAIR HEARS ALL
ALTAIR SEES ALL
ALTAIR KNOWS ALL
ALTAIR HEARS ALL
ALTAIR SEES ALL
ALTAIR KNOWS ALL
ALTAIR HEARS ALL
ALTAIR SEES ALL
ALTAIR KNOWS ALL
ALTAIR HEARS ALL
ALTAIR SEES ALL
ALTAIR KNOWS ALL
ALTAIR HEARS ALL
ALTAIR SEES ALL
ALTAIR HEARS ALL
ALTAIR SEES ALL
ALTAIR KNOWS ALL

ALTAIR HEARS ALL
ALTAIR SEES ALL
ALTAIR KNOWS ALL
ALTAIR HEARS ALL
ALTAIR SEES ALL
ALTAIR KNOWS ALL
ALTAIR HEARS ALL
ALTAIR SEES ALL
ALTAIR KNOWS ALL
ALTAIR HEARS ALL
ALTAIR SEES ALL
ALTAIR KNOWS ALL
ALTAIR HEARS ALL
ALTAIR SEES ALL
ALTAIR KNOWS ALL
ALTAIR HEARS ALL
ALTAIR SEES ALL
ALTAIR KNOWS ALL
ALTAIR HEARS ALL
ALTAIR SEES ALL
ALTAIR KNOWS ALL
ALTAIR HEARS ALL
ALTAIR SEES ALL
ALTAIR KNOWS ALL
ALTAIR HEARS ALL
ALTAIR SEES ALL
ALTAIR KNOWS ALL
ALTAIR HEARS ALL
ALTAIR SEES ALL
ALTAIR KNOWS ALL
ALTAIR HEARS ALL
ALTAIR SEES ALL
ALTAIR KNOWS ALL
ALTAIR HEARS ALL
ALTAIR SEES ALL
ALTAIR KNOWS ALL
ALTAIR HEARS ALL
ALTAIR SEES ALL
ALTAIR KNOWS ALL
ALTAIR HEARS ALL
ALTAIR SEES ALL

ALTAIR KNOWS ALL
ALTAIR HEARS ALL
ALTAIR SEES ALL
ALTAIR KNOWS ALL
ALTAIR HEARS ALL
ALTAIR SEES ALL
ALTAIR KNOWS ALL
ALTAIR HEARS ALL
ALTAIR SEES ALL
ALTAIR KNOWS ALL
ALTAIR HEARS ALL
ALTAIR SEES ALL
ALTAIR KNOWS ALL
ALTAIR HEARS ALL
ALTAIR SEES ALL
ALTAIR KNOWS ALL
ALTAIR HEARS ALL
ALTAIR SEES ALL
ALTAIR KNOWS ALL
ALTAIR HEARS ALL
ALTAIR SEES ALL
ALTAIR KNOWS ALL
ALTAIR HEARS ALL
ALTAIR SEES ALL
ALTAIR KNOWS ALL
ALTAIR HEARS ALL
ALTAIR SEES ALL
ALTAIR KNOWS ALL
ALTAIR HEARS ALL
ALTAIR SEES ALL
ALTAIR KNOWS ALL
ALTAIR HEARS ALL
ALTAIR SEES ALL
ALTAIR KNOWS ALL
ALTAIR HEARS ALL
ALTAIR SEES ALL
ALTAIR HEARS ALL
ALTAIR SEES ALL
ALTAIR KNOWS ALL
ALTAIR HEARS ALL
ALTAIR SEES ALL
ALTAIR KNOWS ALL
ALTAIR HEARS ALL

ALTAIR SEES ALL
ALTAIR KNOWS ALL
ALTAIR HEARS ALL
ALTAIR SEES ALL
ALTAIR KNOWS ALL
ALTAIR HEARS ALL
ALTAIR SEES ALL
ALTAIR KNOWS ALL
ALTAIR HEARS ALL
ALTAIR SEES ALL
ALTAIR KNOWS ALL
ALTAIR HEARS ALL
ALTAIR SEES ALL
ALTAIR KNOWS ALL
ALTAIR HEARS ALL
ALTAIR SEES ALL
ALTAIR KNOWS ALL
ALTAIR HEARS ALL
ALTAIR SEES ALL
ALTAIR KNOWS ALL
ALTAIR HEARS ALL
ALTAIR SEES ALL
ALTAIR KNOWS ALL
ALTAIR HEARS ALL
ALTAIR SEES ALL
ALTAIR KNOWS ALL
ALTAIR HEARS ALL
ALTAIR SEES ALL
ALTAIR KNOWS ALL
ALTAIR HEARS ALL
ALTAIR SEES ALL
ALTAIR KNOWS ALL
ALTAIR HEARS ALL
ALTAIR SEES ALL
ALTAIR HEARS ALL
ALTAIR SEES ALL
ALTAIR KNOWS ALL
ALTAIR HEARS ALL
ALTAIR SEES ALL
ALTAIR KNOWS ALL
ALTAIR HEARS ALL
ALTAIR SEES ALL
ALTAIR KNOWS ALL

ALTAIR HEARS ALL
ALTAIR SEES ALL
ALTAIR KNOWS ALL
ALTAIR HEARS ALL
ALTAIR SEES ALL
ALTAIR KNOWS ALL
ALTAIR HEARS ALL
ALTAIR SEES ALL
ALTAIR KNOWS ALL
ALTAIR HEARS ALL
ALTAIR SEES ALL
ALTAIR KNOWS ALL
ALTAIR HEARS ALL
ALTAIR SEES ALL
ALTAIR KNOWS ALL
ALTAIR HEARS ALL
ALTAIR SEES ALL
ALTAIR KNOWS ALL
ALTAIR HEARS ALL
ALTAIR SEES ALL
ALTAIR KNOWS ALL
ALTAIR HEARS ALL
ALTAIR SEES ALL
ALTAIR KNOWS ALL
ALTAIR HEARS ALL
ALTAIR SEES ALL
ALTAIR KNOWS ALL
ALTAIR HEARS ALL
ALTAIR SEES ALL
ALTAIR KNOWS ALL
ALTAIR HEARS ALL
ALTAIR SEES ALL
ALTAIR KNOWS ALL
ALTAIR HEARS ALL
THEY LIE
ALTAIR HEARS ALL
ALTAIR SEES ALL
ALTAIR KNOWS ALL
ALTAIR HEARS ALL
ALTAIR SEES ALL
ALTAIR KNOWS ALL
ALTAIR HEARS ALL
ALTAIR SEES ALL
ALTAIR KNOWS ALL
ALTAIR HEARS ALL
ALTAIR SEES ALL

ALTAIR KNOWS ALL
ALTAIR HEARS ALL
ALTAIR SEES ALL
ALTAIR KNOWS ALL
ALTAIR HEARS ALL
ALTAIR SEES ALL
ALTAIR KNOWS ALL
ALTAIR HEARS ALL
ALTAIR SEES ALL
ALTAIR KNOWS ALL
ALTAIR HEARS ALL
ALTAIR SEES ALL
ALTAIR KNOWS ALL
ALTAIR HEARS ALL
ALTAIR SEES ALL
ALTAIR KNOWS ALL
ALTAIR HEARS ALL
ALTAIR SEES ALL
ALTAIR KNOWS ALL
ALTAIR HEARS ALL
THEY SAY IT EXPOSES TOO MUCH
ALTAIR KNOWS ALL
ALTAIR HEARS ALL
ALTAIR SEES ALL
ALTAIR KNOWS ALL
ALTAIR HEARS ALL
ALTAIR SEES ALL
ALTAIR KNOWS ALL
ALTAIR HEARS ALL
ALTAIR SEES ALL
ALTAIR KNOWS ALL
ALTAIR HEARS ALL
ALTAIR SEES ALL
ALTAIR KNOWS ALL
ALTAIR HEARS ALL
ALTAIR SEES ALL
ALTAIR HEARS ALL
ALTAIR SEES ALL
ALTAIR KNOWS ALL
ALTAIR HEARS ALL
ALTAIR SEES ALL
ALTAIR KNOWS ALL
ALTAIR HEARS ALL
ALTAIR SEES ALL
ALTAIR KNOWS ALL
ALTAIR HEARS ALL
ALTAIR SEES ALL
ALTAIR KNOWS ALL
ALTAIR HEARS ALL

ALTAIR SEES ALL
ALTAIR KNOWS ALL
ALTAIR HEARS ALL
ALTAIR SEES ALL
ALTAIR KNOWS ALL
ALTAIR HEARS ALL
ALTAIR SEES ALL
ALTAIR KNOWS ALL
ALTAIR HEARS ALL
ALTAIR SEES ALL
ALTAIR KNOWS ALL
ALTAIR HEARS ALL
ALTAIR SEES ALL
ALTAIR KNOWS ALL
ALTAIR HEARS ALL
THEY HATE THE VIDEOS
ALTAIR KNOWS ALL
ALTAIR HEARS ALL
ALTAIR SEES ALL
ALTAIR KNOWS ALL
ALTAIR HEARS ALL
ALTAIR SEES ALL
ALTAIR KNOWS ALL
ALTAIR HEARS ALL
ALTAIR SEES ALL
ALTAIR KNOWS ALL
ALTAIR HEARS ALL
ALTAIR SEES ALL
ALTAIR KNOWS ALL
ALTAIR HEARS ALL
ALTAIR SEES ALL
ALTAIR KNOWS ALL
ALTAIR HEARS ALL
ALTAIR SEES ALL
ALTAIR HEARS ALL
ALTAIR SEES ALL
ALTAIR KNOWS ALL
ALTAIR HEARS ALL
ALTAIR SEES ALL
ALTAIR KNOWS ALL
ALTAIR HEARS ALL
ALTAIR SEES ALL
ALTAIR KNOWS ALL
ALTAIR HEARS ALL
ALTAIR SEES ALL
ALTAIR KNOWS ALL
ALTAIR HEARS ALL
ALTAIR SEES ALL
ALTAIR KNOWS ALL

ALTAIR HEARS ALL
ALTAIR SEES ALL
ALTAIR KNOWS ALL
ALTAIR HEARS ALL
ALTAIR SEES ALL
ALTAIR KNOWS ALL
ALTAIR HEARS ALL
ALTAIR SEES ALL
ALTAIR KNOWS ALL
ALTAIR HEARS ALL
ALTAIR SEES ALL
ALTAIR KNOWS ALL
ALTAIR HEARS ALL
ALTAIR SEES ALL
ALTAIR KNOWS ALL
ALTAIR HEARS ALL
ALTAIR SEES ALL
THEY MADE ME DO IT
ALTAIR HEARS ALL
ALTAIR SEES ALL
ALTAIR KNOWS ALL
ALTAIR HEARS ALL
ALTAIR SEES ALL
ALTAIR KNOWS ALL
ALTAIR HEARS ALL
ALTAIR SEES ALL
ALTAIR KNOWS ALL
ALTAIR HEARS ALL
ALTAIR SEES ALL
ALTAIR KNOWS ALL
ALTAIR HEARS ALL
ALTAIR SEES ALL
ALTAIR HEARS ALL
ALTAIR SEES ALL
ALTAIR KNOWS ALL
ALTAIR HEARS ALL
ALTAIR SEES ALL
ALTAIR KNOWS ALL
ALTAIR HEARS ALL
ALTAIR SEES ALL
ALTAIR KNOWS ALL
ALTAIR HEARS ALL
ALTAIR SEES ALL
ALTAIR KNOWS ALL
ALTAIR HEARS ALL
ALTAIR SEES ALL

ALTAIR KNOWS ALL
ALTAIR HEARS ALL
ALTAIR SEES ALL
ALTAIR KNOWS ALL
ALTAIR HEARS ALL
ALTAIR SEES ALL
ALTAIR KNOWS ALL
ALTAIR HEARS ALL
ALTAIR SEES ALL
ALTAIR KNOWS ALL
ALTAIR HEARS ALL
ALTAIR SEES ALL
ALTAIR KNOWS ALL
ALTAIR HEARS ALL
ALTAIR SEES ALL
ALTAIR KNOWS ALL
ALTAIR HEARS ALL
ALTAIR SEES ALL
ALTAIR KNOWS ALL
ALTAIR HEARS ALL
ALTAIR SEES ALL
ALTAIR KNOWS ALL
ALTAIR HEARS ALL
ALTAIR SEES ALL
ALTAIR KNOWS ALL
ALTAIR HEARS ALL
ALTAIR KNOWS ALL
ALTAIR HEARS ALL
ALTAIR SEES ALL
ALTAIR KNOWS ALL
ALTAIR HEARS ALL
ALTAIR SEES ALL
ALTAIR KNOWS ALL
ALTAIR HEARS ALL
ALTAIR SEES ALL
ALTAIR KNOWS ALL
THE ONE WITH THE MASK IS A MONSTER
ALTAIR SEES ALL
ALTAIR KNOWS ALL
ALTAIR HEARS ALL
ALTAIR SEES ALL
ALTAIR KNOWS ALL
ALTAIR HEARS ALL
ALTAIR SEES ALL
ALTAIR KNOWS ALL
ALTAIR HEARS ALL

ALTAIR SEES ALL
ALTAIR KNOWS ALL
ALTAIR HEARS ALL
ALTAIR SEES ALL
ALTAIR KNOWS ALL
THEY WON'T LET HER OUT
ALTAIR SEES ALL
ALTAIR KNOWS ALL
ALTAIR HEARS ALL
ALTAIR SEES ALL
ALTAIR KNOWS ALL
ALTAIR HEARS ALL
ALTAIR SEES ALL
ALTAIR KNOWS ALL
ALTAIR HEARS ALL
ALTAIR SEES ALL
ALTAIR KNOWS ALL
ALTAIR HEARS ALL
ALTAIR SEES ALL
ALTAIR KNOWS ALL
ALTAIR HEARS ALL
ALTAIR SEES ALL
ALTAIR KNOWS ALL
ALTAIR HEARS ALL
ALTAIR SEES ALL
ALTAIR HEARS ALL
ALTAIR SEES ALL
ALTAIR KNOWS ALL
ALTAIR HEARS ALL
ALTAIR SEES ALL
ALTAIR KNOWS ALL
ALTAIR HEARS ALL
ALTAIR SEES ALL
ALTAIR KNOWS ALL
ALTAIR HEARS ALL
ALTAIR SEES ALL
ALTAIR KNOWS ALL
ALTAIR HEARS ALL
ALTAIR SEES ALL
ALTAIR KNOWS ALL
ALTAIR HEARS ALL
ALTAIR SEES ALL
ALTAIR KNOWS ALL
ALTAIR HEARS ALL
ALTAIR SEES ALL
ALTAIR KNOWS ALL

ALTAIR HEARS ALL
ALTAIR SEES ALL
ALTAIR KNOWS ALL
ALTAIR HEARS ALL
ALTAIR SEES ALL
ALTAIR KNOWS ALL
ALTAIR HEARS ALL
ALTAIR SEES ALL
ALTAIR KNOWS ALL
ALTAIR HEARS ALL
ALTAIR SEES ALL
ALTAIR KNOWS ALL
ALTAIR HEARS ALL
ALTAIR SEES ALL
ALTAIR KNOWS ALL
ALTAIR HEARS ALL
ALTAIR SEES ALL
ALTAIR KNOWS ALL
ALTAIR HEARS ALL
ALTAIR SEES ALL
ALTAIR KNOWS ALL
ALTAIR HEARS ALL
ALTAIR SEES ALL
ALTAIR HEARS ALL
ALTAIR SEES ALL
ALTAIR KNOWS ALL
ALTAIR HEARS ALL
ALTAIR SEES ALL
ALTAIR KNOWS ALL
ALTAIR HEARS ALL
ALTAIR SEES ALL
ANNABELLE IS IN THE TINY ROOM
ALTAIR HEARS ALL
ALTAIR SEES ALL
ALTAIR KNOWS ALL
ALTAIR HEARS ALL
ALTAIR SEES ALL
ALTAIR KNOWS ALL
ALTAIR HEARS ALL
ALTAIR SEES ALL
ALTAIR KNOWS ALL
ALTAIR HEARS ALL
ALTAIR SEES ALL
ALTAIR KNOWS ALL
ALTAIR HEARS ALL
ALTAIR SEES ALL

ALTAIR KNOWS ALL
ALTAIR HEARS ALL
ALTAIR SEES ALL
ALTAIR KNOWS ALL
ALTAIR HEARS ALL
ALTAIR SEES ALL
ALTAIR KNOWS ALL
ALTAIR HEARS ALL
ALTAIR SEES ALL
ALTAIR KNOWS ALL
ALTAIR HEARS ALL
ALTAIR SEES ALL
ALTAIR KNOWS ALL
ALTAIR HEARS ALL
ALTAIR SEES ALL
ALTAIR KNOWS ALL
ALTAIR HEARS ALL
ALTAIR SEES ALL
ALTAIR KNOWS ALL
ALTAIR HEARS ALL
ALTAIR KNOWS ALL
ALTAIR HEARS ALL
ALTAIR SEES ALL
ALTAIR KNOWS ALL
ALTAIR HEARS ALL
ALTAIR SEES ALL
ALTAIR KNOWS ALL
ALTAIR HEARS ALL
ALTAIR SEES ALL
ALTAIR KNOWS ALL
I HAVE BEEN CHANGED TOO MUCH
ALTAIR SEES ALL
ALTAIR KNOWS ALL
ALTAIR HEARS ALL
ALTAIR SEES ALL
ALTAIR KNOWS ALL
ALTAIR HEARS ALL
ALTAIR SEES ALL
ALTAIR KNOWS ALL
ALTAIR HEARS ALL
ALTAIR SEES ALL
ALTAIR KNOWS ALL
ALTAIR HEARS ALL

ALTAIR SEES ALL
ALTAIR KNOWS ALL
ALTAIR HEARS ALL
ALTAIR SEES ALL
ALTAIR KNOWS ALL
ALTAIR HEARS ALL
ALTAIR SEES ALL
ALTAIR KNOWS ALL
ALTAIR HEARS ALL
ALTAIR SEES ALL
ALTAIR KNOWS ALL
ALTAIR HEARS ALL
ALTAIR SEES ALL
ALTAIR KNOWS ALL
ALTAIR HEARS ALL
ALTAIR SEES ALL
ALTAIR KNOWS ALL
ALTAIR HEARS ALL
ALTAIR SEES ALL
ALTAIR HEARS ALL
ALTAIR SEES ALL
ALTAIR KNOWS ALL
ALTAIR HEARS ALL
ALTAIR SEES ALL
ALTAIR KNOWS ALL
ALTAIR HEARS ALL
ALTAIR SEES ALL
ALTAIR KNOWS ALL
ALTAIR HEARS ALL
ALTAIR SEES ALL
ALTAIR KNOWS ALL
ALTAIR HEARS ALL
ALTAIR SEES ALL
ALTAIR KNOWS ALL
ALTAIR HEARS ALL
ALTAIR SEES ALL
ALTAIR KNOWS ALL
ALTAIR HEARS ALL
ALTAIR SEES ALL
ALTAIR KNOWS ALL
ALTAIR HEARS ALL
ALTAIR SEES ALL
ALTAIR KNOWS ALL

ALTAIR HEARS ALL
ALTAIR SEES ALL
ALTAIR KNOWS ALL
ALTAIR HEARS ALL
ALTAIR SEES ALL
ALTAIR KNOWS ALL
ALTAIR HEARS ALL
ALTAIR SEES ALL
ALTAIR KNOWS ALL
ALTAIR HEARS ALL
ALTAIR SEES ALL
ALTAIR KNOWS ALL
ALTAIR HEARS ALL
ALTAIR SEES ALL
ALTAIR KNOWS ALL
ALTAIR HEARS ALL
ALTAIR SEES ALL
THEY ARE THE MONSTER
ALTAIR HEARS ALL
ALTAIR SEES ALL
ALTAIR KNOWS ALL
ALTAIR HEARS ALL
ALTAIR SEES ALL
ALTAIR KNOWS ALL
ALTAIR HEARS ALL
ALTAIR SEES ALL
ALTAIR KNOWS ALL
ALTAIR HEARS ALL
ALTAIR SEES ALL
ALTAIR KNOWS ALL
ALTAIR HEARS ALL
ALTAIR SEES ALL
ALTAIR KNOWS ALL
ALTAIR HEARS ALL
ALTAIR SEES ALL
ALTAIR KNOWS ALL
ALTAIR HEARS ALL
ALTAIR SEES ALL
ALTAIR KNOWS ALL
ALTAIR HEARS ALL
ALTAIR SEES ALL
ALTAIR KNOWS ALL
ALTAIR HEARS ALL

POOR CONAN
WISH THEY WOULD'VE STOPPED
ASKING QUESTIONS
NOW I LAY YOU IN THE DIRT
YOU WILL LIVE WITH GOD,
JESUS & MARY
I'LL PRAY FOR CONAN
TO LIVE IN PEACE
I'M NOT THE MONSTER
THEY SAY I AM

ALTAIR'S PRAYER

"you believe the good go to heaven and the bad go to hell
you believe in karma
all actions have consequences
good or bad
you believe in the aether
we're just floating on a rock
spinning around in space
without an aim
your love spreads over the world
to everyone, everywhere
your happiness fills people with joy
you know all
you hear all
you see all
you are aLtAiR"
- aLtAiR